LOGANSPORT CASS COUNTY PUBLIC LIBRARY

AN OUTDOOR SCIENCE BOOK

RAINDROPS ❖❖❖ AND ❖❖❖ RAINBOWS

ROSE WYLER
PICTURES BY STEVEN JAMES PETRUCCIO

JULIAN Ⓜ MESSNER

Text copyright © 1989 by Rose Wyler
Illustrations copyright © 1989 by
Steven James Petruccio.
All rights reserved including the right
of reproduction in whole or in part in
any form. Published by Julian Messner,
a division of Silver Burdett Press, Inc.,
Simon & Schuster, Inc. Prentice Hall
Bldg., Englewood Cliffs, NJ 07632.

JULIAN MESSNER and colophon are
trademarks of Simon & Schuster, Inc.
Design by Malle N. Whitaker.
Manufactured in the United States of
America.

(lib.)   10 9 8 7 6 5 4 3 2 1
(pbk.) 10 9 8 7 6 5 4 3 2

Library of Congress Cataloging-in-
Publication Data
Wyler, Rose.
Raindrops and Rainbows/Rose Wyler;
pictures by Steven James Petruccio.
    p.   cm.—(An Outdoor science book)
Summary: Simple text and experiments illustrate why it rains, how
clouds and rainbows form, and why
lightning and thunder occurs.
1. Rain and rainfall—Juvenile liter-
ature. 2. Rain and rainfall—Experi-
ments—Juvenile literature. 3. Rainbow
—Juvenile literature. 4. Rainbow—
Experiments—Juvenile literature.
5. Clouds—Juvenile literature. 6. Clouds
—Experiments—Juvenile literature.
[1. Rain and rainfall. 2. Clouds.
3. Weather.] I. Petruccio, Steven
James, ill. II. Title. III. Series: Wyler,
Rose. Outdoor science book.
QC924.7.W95 1989   551.57—dc19
88-31191
CIP   AC
ISBN 0-671-66346-1 (lib. bdg.)
ISBN 0-671-66350-X (pbk.)

The author and publisher thank
Donald Witten, of the National
Weather Service, for his helpful
suggestions.

What Makes It Rain?

Down comes the rain, pouring from the sky.
The raindrops patter,
splash and splatter.
Puddles grow and streams fill up.

You wonder where all the water comes from.
How does it get up into the sky?
Why don't you see it go up?

Water is not always wet.
Sometimes it dries up and goes into the air
in bits too small to see or feel.
The bits come from seas, lakes and rivers —
from puddles too and from wash hung out to dry.
Although you cannot see these bits of water,
you can find out how they get into the air.

◆ ◆ ◆ ◆

First fill two small glasses with water. Then add a drop of food coloring to each glass. Keep one in the sun and one in the shade. The water in the shade stays cool for hours and the glass stays almost full. But most of the water in the sun dries up. As it gets warm, tiny bits go into the air. There they form a gas that you cannot see.

Whenever water is warmed by the sun, bits of it form an invisible gas.

High in the air, some of the invisible gas suddenly turns into a beautiful cloud. For a while the cloud floats in a blue sky. Then as the warm, moist cloud rises, it cools. It can't hold as much water, and it breaks into millions of silvery raindrops.

Sounds like a fairy tale, doesn't it? Yet rain really starts that way.

To see how a cloud forms, try making one. Rinse an empty bottle with warm water. Then pour out all but an inch of the water. Take an empty metal tray from the freezer. Place the cold tray on top of the bottle and suddenly a cloud of water droplets forms in it.

The warm water turns into gas and then the gas cools and forms a cloud of tiny droplets. But no rain falls from your homemade cloud. The cloud is not cold enough.

◆◆◆◆

To get rain from a homemade cloud, try this. Ask a grownup to boil some water in a teakettle, and watch the cloud that rises above it. The cloud is warm, so no rain falls from it.

Now set an empty tray below the cloud. Hold a tray of ice cubes above it. The cloud gets cold, and down comes the rain.

Outdoors, most clouds do not become rainmakers.
Raindrops form only in very cold clouds.
Little drops run into other drops and
get bigger and bigger until finally they fall.

Watch raindrops on a window and you see
that they get bigger too
as they run into other drops.

Sometimes tall, puffy clouds become rain clouds.
A shower starts and big drops fall.
They splatter and splash, but not for long.
The shower usually ends quickly.

Small raindrops come from gray clouds
that spread evenly across the sky.
This kind of cloud is not very tall,
but it often stretches for miles and miles
and the rain may last for days.

To check the size of raindrops, catch some on a jar lid that is covered with flour. Bring the lid indoors and inspect the doughballs that were made by the raindrops.

Take another jar lid covered with flour and squirt drops of water on it with an eyedropper. The doughballs they form are about as big as the ones that most raindrops make. Did the rain make bigger or smaller ones?

How much rain comes down in a storm? To find out, put a jar outdoors. When the rain stops, hold a ruler against the jar and see how deep the water is. Water that deep fell wherever it rained.

How many drops of water can you get out of the jar with an eyedropper? Even if the jar is small, you may get hundreds. Just think how many drops fell during the storm. Millions, and billions, and trillions!

Lightning, then Thunder

Sometimes on a hot, sticky day,
a puffy cloud gets puffier and taller.

When it reaches very cold air,
the top of the cloud flattens out.
Now it is thousands of feet tall.

The cloud gets wider and darker.
Then flash, lightning! Crash, thunder!
Rain pours down, slashing the ground.

Winds rage in the storm cloud.
Raindrops get tossed around
and rub against each other.
Electricity builds up on them.
Suddenly the electricity turns
into a huge hot spark of lightning.
The spark rips through the air,
shoving cold air aside.
Great waves of sound start and
thunder roars and rolls.
More lightning forms and flashes
until finally the winds die down.

Sometimes you see lightning from a faraway cloud quite a while before you hear thunder.
That's because sound moves more slowly than light. It moves about a mile in 5 seconds.

To tell how near a storm is, count the seconds between the lightning and thunder. Since it takes 1 second to say one thousand one, just count this way: 1001, 1002, 1003 and so on.

Thunderstorms travel quite fast—
about half a mile a minute.
One that is 2 miles away and coming closer
will arrive in 4 minutes.

Use that time to go to a safe place—
inside a car or a building.
Then watch the wonderful fireworks in the sky.
The flashes and crashes will stop soon,
for most thunderstorms end in half an hour.

Looking for Rainbows

If the sun comes out before the storm ends, turn your back to it and look for a rainbow.

You see a rainbow as sunlight passes through the raindrops falling in front of you.
No one else can look through those same drops
So no one else sees the same rainbow you see, not even someone who stands next to you.

◆◆◆◆
You do not need rain to have a rainbow. You can make one any morning or afternoon when the sun is low in the sky. Just turn on a garden hose. Stand with your back to the sun, and you see a beautiful rainbow.

Red, orange, yellow, green, blue, purple — all those rainbow colors are in sunlight. They do not show for they are mixed together. But if sunlight hits drops of water at a slant, the colors separate and the drops reflect them. Then you see each rainbow color.

◆◆◆◆

Try this too while the sun is shining. Set a mirror at a slant in a pan full of water. Hold it so that sunlight hits it, then is reflected on a white wall. Presto! the rainbow colors in sunlight separate and each one forms a band on the wall.

To make the bands vanish, stir the water. The colors mix and as sunlight leaves the water, it is white, its usual color.

◆◆◆◆

Here's another way to mix the rainbow colors.

1. Cut a circle 4 inches wide from white cardboard.
2. Color it like the wheel in the picture.
3. Near the center, punch two holes with a pencil. Thread 1 yard of string through the holes.
4. Tie the ends together, then set the wheel in the middle as shown.

◆◆◆◆

Everything is ready now.
5. Hold the string by the ends and twirl the wheel. When the string is tightly twisted, pull it.
6. The wheel spins, the colors mix. The colors turn pale gray — almost white. If they were pure, they would turn pure white.

Sometimes you see a ring of rainbow colors.
The ring forms around the full moon
when it shines through high clouds of haze.
Moonlight, like sunlight, has the rainbow colors
in it, all mixed together.
As it passes through haze, the colors separate.
Since haze clouds are made of ice needles,
the ring is really an ice bow—not a rainbow.

Clouds Come and Go

"A ring around the moon
means rain is coming soon."

So people say — and they are often right.
High haze clouds are often followed by
low gray clouds made of water droplets.
Overnight the droplets grow into raindrops.
The clouds thicken and darken — then rain start

Clouds of haze in daytime are like haze at night
They too warn that rain is coming soon.

25

You can also tell the weather will change when you see thin, curly wisps of clouds. High winds soon blow these clouds away. Then you see rows of small, shaggy clouds that look a little like scales on a fish. Fishermen call this a mackerel sky, saying, "Mackerel sky—sometimes wet, sometimes dry." And while it lasts the weather is fair.

Some time later
clouds form in sheets.
They get thick and dark.
Then there is rain.

If it is very cold,
water freezes in the clouds.
Then there may be snow.

You hear the patter of raindrops.
But snow falls without making a sound.
In a storm, tons and tons of snow fall silently.

If you can, catch some snow on black paper.
Close up, the flakes look like fine lace.
They glisten because they are made of ice.
Each is different, but each has six sides.
Air fills the holes in the icy lace
and that's why snow falls silently.

Snow is really a frozen water fluff—
it has so much air in it.
if you melt a cupful of snow, you get
about one-tenth of a cup of water.

After a day or two, wet weather ends in most parts of the United States and Canada. West winds bring fresh, dry air and drive trains of puffy clouds across the sky.

The puffy clouds are so thick that sunlight cannot go through them.
When they pass in front of the sun, they look dark and cast shadows.
But it will not rain unless the clouds grow tall.

Clouds are a kind of sky writing.
They bring weather news.
Right now, what is the sky like?
Tomorrow, will there be rain—
and perhaps a rainbow?